SAN HEKAÝASY

THE NUMBER STORY

SMALL BOOK ONE

ENGLISH - TURKMEN

Numbers Teach Children
Their Number Names

written and illustrated by

MISS ANNA

Early Reader Edition of *The Number Story 1*
Bronze Medal Winner, 2016 Wishing Shelf Book Award

Library of Congress Control Number: 2018902040

Names: Miss Anna, author.
Title: Number story : numbers teach children their number names / Miss Anna.
Description: Portland, OR: Lumpy Publishing, 2018.
Identifiers: ISBN 978-1-949320-00-8| LCCN 2018902040
Summary: The pictures and rhymes present stories which introduce numbers 0-10.
Subjects: LCSH Numeration—English—Turkmen--Pictorial works--Juvenile literature. | BISAC JUVENILE NONFICTION /
Languages: English—Turkmen
Classification: LCC QA141.3 .M57 2018 | DDC 513—dc23

Publisher: Lumpy Publishing
Website: www.missannabooks.com
Email: missanna@missannabooks.com

Paperback: ISBN 978-1-949320-00-8
Printed in the U.S.A.   1 3 5 7 9 10 8 6 4 2

Sanlaryň atlaryny öwrenmek islärmidiň?

It is very easy and a lot of fun!

Ol gaty aňsat we onda şatlyk köp!

Say-along our little jingle

Biz bilen kiçijek hekaýamyzyň

aýdymyna goşul!

starting from Number One!

Bir Sanyndan başlaýarys!

1

ONE looks like my one finger.

BIR

Bir meniň bir barmagyma meňzeýär.

1
ONE!
BIR!

2
TWO trails a tail.
IKI
Iki guýruk çekýär.

A TAIL! GUÝRUK!

3

THREE has bumps.

ÜÇ

Üçüň pökgülikleri bar.

BUMPY! PÖKGI!

4

FOUR carries a sail.

DÖRT

Dört ýelken çekýär.

4
A SAIL!
ÝELKEN!

FIVE is a racing track.

BÄŞ

Bäş ýaryş tregidir.

VROOM
WRUUM!

S I X   curves like a snail.

ALTY

Alty balykgulak ýaly egrelýär.

A SNAIL!   BALYKGULAK!

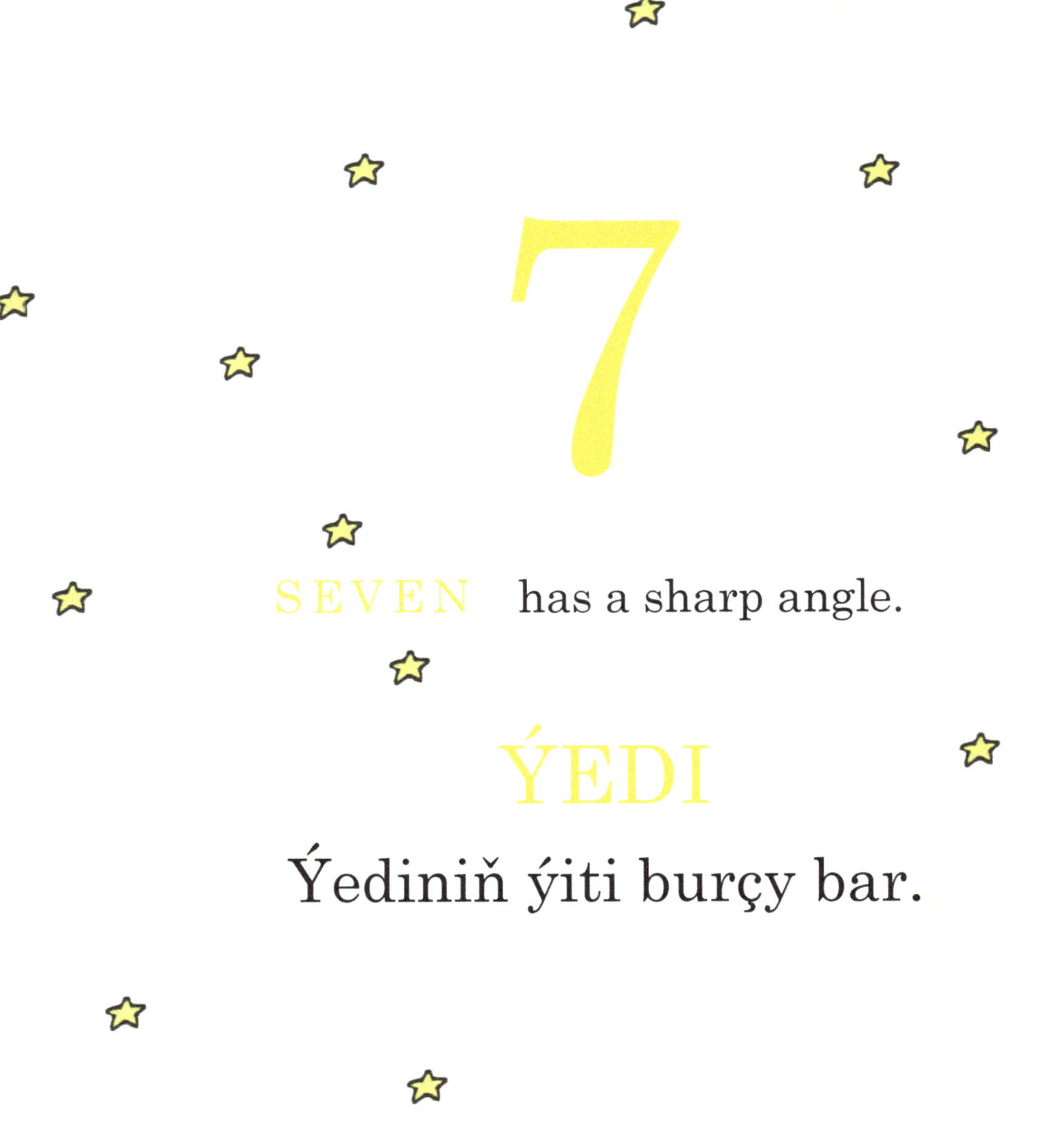

7

S E V E N has a sharp angle.

ÝEDI

Ýediniň ýiti burçy bar.

BE CAREFUL! IT'S SHARP!

Seresap bol! Ol ýiti!

SEKIZ

Sekiz gorkanyň relsleridir.

URA!
YIPPEE!

NINE   is a bubble on a stick.

DOKUZ

Dokuz taýajykdaky köpürjikdir.

A BUBBLE!

KÖPÜRJIK!

10

TEN   is an eye of a whale.

ON

On läheňiň bir gözüdir.

WINK!
GÖZ GYRP!
HELLO! SALAM!

And
We

O

ZERO   is an empty pail.

NOL

Nol boş bedredir.

IT'S EMPTY!
Ol Boş!

Thank you for playing with us today.

We had a lot of fun too!

Şugün biziň bilen oýnanyň üçin minnetdar.

Biz hem köp şatlandyk!

We are your Number friends,
Zero to Ten,
Who will be here for you~
Biz seniň San dostlaryň
Noldan Ona.
Biz hemişe seniň üçin şu ýerde bolarys!

Bye-bye now!
See you again soon!
Häzir bolsa sagbol!
Ýakynda görüşeris!

The Numbers are *SINGING* too!

To sing-a-long, look for Miss Anna Number Story
at your favorite music store like iTUNES.

MP3

Numbers 0-10
IDENTIFYING & COUNTING

Numbers 11-20
& Ordinals
first, second, third…

Numbers 0-100
& Place Values
ones, tens, hundreds…

About Clocks
& Telling Time
hours, minutes, seconds

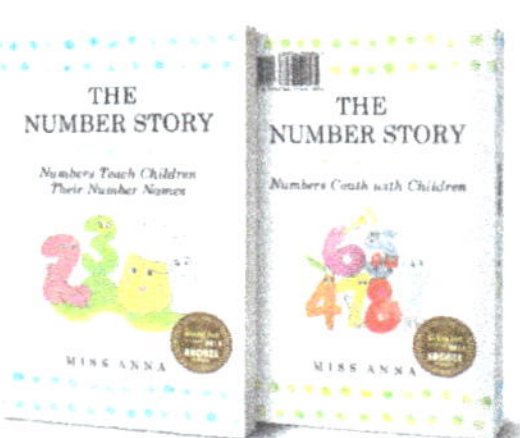

Number Story 1 & 2
isbn: 978-0-996216-48-7

Number Story 3 & 4
isbn: 978-1-945977-01-5

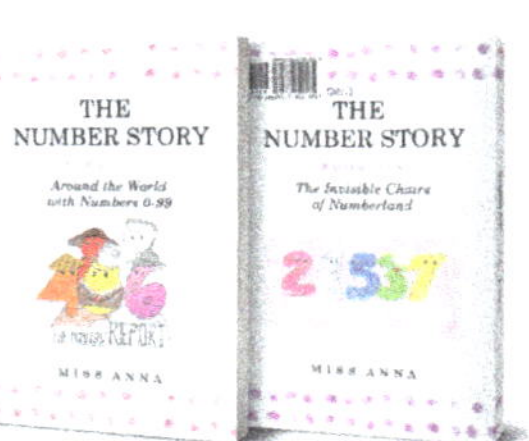

Number Story 5 & 6
isbn: 978-1-945977-06-0

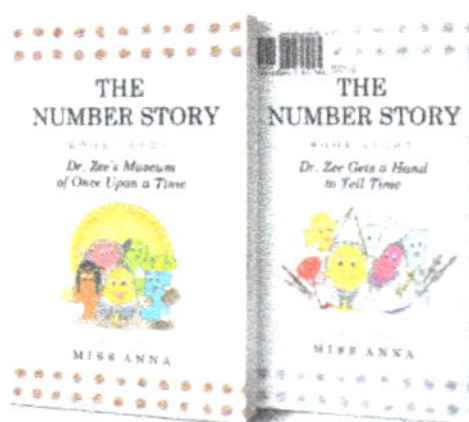

Number Story 7 & 8
isbn: 978-1-949320-40-

For more Miss Anna books to love,
visit us at

www.missannabooks.com

Numbers are working hard all over the world!
Come Travel the World with Us!